February 2011

CLIMATE CHANGE ISSUES

Options for Addressing Challenges to Carbon Offset Quality

GAO
Accountability • Integrity • Reliability

Highlights

Highlights of GAO-11-345, a report to the Chairman, Committee on Oversight and Government Reform, House of Representatives

Why GAO Did This Study

Carbon offsets are reductions in greenhouse gas emissions in one place to compensate for emissions elsewhere. Examples of offset projects include planting trees, developing renewable energy sources, or capturing emissions from landfills. Recent congressional proposals would have limited emissions from utilities, industries, or other "regulated entities," and allowed these entities to buy offsets. Research suggests that offsets can significantly lower the cost of a program to limit emissions because buying offsets may cost regulated entities less than making the reductions themselves.

Some existing international and U.S. regional programs allow offsets to be used for compliance with emissions limits. A number of voluntary offset programs also exist, where buyers do not face legal requirements but may buy offsets for other reasons. Prior GAO work found that it can be difficult to ensure offset quality—that offsets achieve intended reductions. One quality criterion is that reductions must be "additional" to what would have occurred without the offset program.

This report provides information on (1) key challenges in assessing the quality of different types of offsets and (2) options for addressing key challenges associated with offset quality if the U.S. adopted a program to limit emissions. GAO reviewed relevant literature and interviewed selected experts and such stakeholders as project developers, verifiers, and program officials. This report contains no recommendations.

View GAO-11-345 or key components. For more information, contact David Trimble at (202) 512-3841 or trimbled@gao.gov.

What GAO Found

According to experts, stakeholders, and available information, key challenges in assessing the quality of offset projects include the following:

- **Additionality.** According to many experts and stakeholders GAO interviewed, additionality is the primary challenge to offset quality. Assessing additionality is difficult because it involves determining what emissions would have been without the incentives provided by the offset program. Studies suggest that existing programs have awarded offsets that were not additional.
- **Measuring and managing soil and forestry offsets.** For projects that store carbon in soils and forests, it is challenging to estimate the amount of carbon stored and to manage the risk that carbon may later be released by, for example, fires or changes in land management. Some studies have estimated that projects involving soils and forestry could constitute the majority of offsets under a U.S. program.
- **Verification.** Experts and stakeholders said that verifying offsets in existing markets has presented several challenges. In particular, project developers and offset buyers may have few incentives to report information accurately or to investigate offset quality.

According to experts, stakeholders, and available information, policymakers have several options to choose from in addressing challenges with offset quality. These approaches often involve fundamental trade-offs, such as increasing the cost of offsets. Nevertheless, some research indicates that including offsets in a program to limit emissions could provide substantial cost savings that would not be provided by a program without offsets.

- **Additionality.** One way to assess additionality is project-by-project approval, a lengthy process that considers the individual circumstances of each project. Another approach is to group projects into categories and apply a standard to the entire group—for example, award offsets to all electricity generators with emissions below a certain level. While such standards may be less subjective and less costly to administer, they may also require a considerable up-front investment to collect data for various project types.
- **Measuring and managing soil and forestry offsets.** To address these challenges a program could, for example, adjust the amount of offsets awarded based on measurement uncertainty, or establish a "buffer pool" of offsets to compensate for any re-released carbon.
- **Verification.** To address this challenge, a program could, for example, hold verifiers liable for problems with offsets they have approved, contract with independent verifiers, and provide for rigorous oversight.

Experts also identified options that could address multiple quality assurance challenges, such as limiting the quantity or type of offsets that can be used for compliance. However, limiting the supply of offsets could also raise their cost. Regardless of the program design, many experts said an offset program should clearly identify goals, align incentives with goals, promote transparency, and continuously evaluate progress.

_____ **United States Government Accountability Office**

Contents

Abbreviations

ANSI	American National Standards Institute
CAR	Climate Action Reserve
CBO	Congressional Budget Office
CCX	Chicago Climate Exchange
CDM	Clean Development Mechanism
DNV	Det Norske Veritas
EPA	Environmental Protection Agency
EU ETS	European Union Emissions Trading System
ISO	International Standards Organization
REDD	Reduced Emissions from Deforestation and Degradation
RGGI	Regional Greenhouse Gas Initiative
UFS	Voluntary Carbon Standard
UNFCCC	United Nations Framework Convention on Climate Change
VCS	Voluntary Carbon Standard

United States Government Accountability Office
Washington, DC 20548

February 15, 2011

The Honorable Darrell Issa
Chairman
Committee on Oversight
    and Government Reform
House of Representatives

Dear Mr. Chairman:

In the past year, Congress has considered proposals to limit greenhouse gas emissions from many sectors of the economy, including electric power generation, transportation, and manufacturing.[1] Most of these proposals have focused on market-based mechanisms such as cap-and-trade, a system the United States already uses to reduce air pollution that causes acid rain. Under a cap-and-trade program, the government would place an overall cap on emissions and issue tradable permits. Entities covered by the program would have to surrender enough permits for all of their emissions at the end of specified time periods. Such market-based programs could forestall some of the potentially adverse effects of climate change at less cost than other options to regulate emissions. However, a program that reduces emissions could also increase the cost of activities that generate emissions, such as the burning of fossil fuels. As a result, cap-and-trade proposals have also included various provisions aimed at limiting costs to businesses and consumers.

One potential cost containment mechanism for a cap-and-trade program is the use of carbon offsets—activities that reduce emissions in one place in order to compensate for emissions occurring elsewhere.[2] Examples of offset projects include (1) planting trees; (2) capturing greenhouse gases from mines, landfills, and agricultural operations; (3) reducing tilling to

---

[1]There are six primary greenhouse gases: carbon dioxide, methane, nitrous oxide, and three synthetic gases—hydrofluorocarbons, perfluorocarbons, and sulfur hexafluoride.

[2]This report uses the term *carbon offsets* to describe offsets derived from any of the six primary greenhouse gases. Carbon offsets are typically quantified and described in terms of metric tons of carbon dioxide equivalent. Carbon dioxide equivalents provide a common standard for measuring the warming potential of different greenhouse gases and are calculated by multiplying the emissions of the non-carbon dioxide gas by its global warming potential, a factor that measures its heat-trapping ability relative to that of carbon dioxide.

store, or "sequester," more carbon in agricultural soil; (4) installing more energy-efficient equipment; and (5) generating renewable energy from hydroelectric, wind, or solar power. Such projects produce tradable credits, or "offsets," which can be purchased by regulated entities and used to comply with emissions caps.[3] In principle, allowing the use of offsets would provide regulated entities with greater flexibility to make emissions reductions at less cost. Regulated entities may find that it is cheaper to reduce emissions by purchasing an offset than it is to reduce their own emissions or to purchase permits from another regulated entity. For example, it may cost less to pay a landfill owner or operator to capture greenhouse gas emissions than to reduce emissions at a power plant.

U.S. legislation proposed in the past year would have created a cap-and-trade program that allowed regulated entities to use offsets to comply with emissions caps, as does the European Union Emissions Trading System (EU ETS) and other existing programs that limit greenhouse gas emissions.[4] Economic research indicates that including offsets in a cap-and-trade program could provide substantial cost savings. For example, in an analysis of the American Clean Energy Security Act, Congressional Budget Office (CBO) estimated that from 2012 through 2050, the annual net cost of a program allowing offsets would be about 70 percent less than a program without offsets.[5] The extent of any savings is uncertain and would depend on many factors, including the design of the regulatory and offset programs. Such decisions could greatly influence an offset market that, under some past legislative proposals, could become many times greater than the largest existing offset market, which involves billions of dollars worth of offset transactions each year.

---

[3]Although carbon offsets have primarily been considered as part of a cap-and-trade proposal, they could be used to limit the costs of a variety of programs to limit greenhouse gas emissions.

[4]The EU ETS, which commenced operation in January 2005, is the world's largest greenhouse gas cap-and-trade program. For more information on the EU ETS, see GAO, *International Climate Change Programs: Lessons Learned from the European Union's Emissions Trading Scheme and the Kyoto Protocols Clean Development Mechanism*, GAO-09-151 (Washington, D.C.: Nov. 18, 2008).

[5]CBO, *The Use of Offsets to Reduce Greenhouse Gases* (Washington, D.C.: Aug 3, 2009). According to CBO, this figure includes an estimate of the costs involved in an offset program, such as administration costs and measures taken to address offset quality, but does not provide insight into whether offsets provide the full intended reductions.

However, we have previously reported that carbon offsets may also compromise the environmental integrity of programs to limit emissions and should therefore be carefully evaluated.[6] Among other things, we identified challenges that can affect the quality of carbon offsets. A quality offset is one that achieves its intended reductions—in most programs, this means that one offset credit equals one ton of reduced or avoided emissions. While definitions vary, our review of the literature points to five general criteria for assessing offset quality—an offset must be additional, real, verifiable, permanent, and enforceable. An offset is *additional* if it would not have occurred without the incentives provided by the offset program. *Real* means that the quantified emissions reductions represent actual net emissions reductions, and are not a product of incomplete or inaccurate accounting; *verifiable* means the reductions associated with the project can be accurately quantified, monitored, and verified; *permanent* means the emissions stored by a project will not be released into the atmosphere in the future, or that there are guarantees to ensure that such releases are replaced; and *enforceable* ensures that offsets are backed by tracking systems that define their ownership as well as regulations and penalties for noncompliance.

Some legislative proposals to limit greenhouse gases, if enacted, would have involved a number of federal agencies in the development of offset quality standards and program oversight. The discussion draft of the 2010 American Power Act, for example, would have given the Environmental Protection Agency (EPA) primary oversight over domestic offsets—except for those pertaining to agriculture and forestry, for which the Department of Agriculture would have had primary responsibility.[7]

This report responds to your request for a review of offset quality issues. This report provides information on (1) the key challenges in assessing the quality of different types of offset projects, and (2) options for addressing key challenges associated with offset quality if the United States adopted a program to limit greenhouse gas emissions. To respond to these

---

[6]See GAO-09-151; GAO, *Carbon Offsets: The U.S. Voluntary Market Is Growing but Quality Assurance Poses Challenges for Market Participants*, GAO-08-1048 (Washington, D.C.: Aug. 29, 2008); and *Climate Change: Observations on the Potential Role of Carbon Offsets in Climate Change Legislation*, Testimony Before the Subcommittee on Energy and Environment, Committee on Energy and Commerce, House of Representatives, GAO-09-456T (Washington, D.C.: Mar. 5. 2009).

[7]American Power Act (discussion draft), available at http://kerry.senate.gov/imo/media/doc/APAbill3.pdf

objectives, we reviewed relevant literature and interviewed 13 experts—including economists, academic researchers, and experts in ecology and law—selected based on their experience, recommendations from persons knowledgeable in climate policy issues, and the relevance and extent of their publications. We also assessed approaches used in seven offset programs selected based on their representation in literature, and interviewed 17 stakeholders—project developers, verifiers, and program officials—from these programs. Information from our sample of experts and stakeholders cannot be generalized to those we did not speak to. Appendix I provides additional information about our scope and methodology, and appendix II lists the experts and stakeholders we interviewed.

We conducted our work from April 2010 to February 2011 in accordance with all sections of GAO's Quality Assurance Framework that are relevant to our objectives. The framework requires that we plan and perform the engagement to obtain sufficient and appropriate evidence to meet our stated objectives and to discuss any limitations in our work. We believe that the information and data obtained, and the analysis conducted, provide a reasonable basis for any findings and conclusions in this product.

Background

Carbon offsets can be used by entities that are subject to legal requirements to limit their emissions, such as utilities or manufacturing facilities. Offset programs designed for this purpose are called compliance programs. One such program is the Clean Development Mechanism (CDM), an offset program established by the Kyoto Protocol.[8] The CDM allows nations with binding emissions targets under the Kyoto Protocol—including those participating in the EU ETS—to purchase offsets from projects in developing nations without binding targets. The CDM is the world's largest offset market, valued at $2.7 billion in 2009, and has registered over 2,700 offset projects in 70 countries.[9] Our prior work found that the CDM provided developed nations with flexibility in meeting their emissions targets but that the program's effects on emissions were

---

[8]The Kyoto Protocol is an international agreement to limit the adverse effects of climate change developed within the United Nations Framework Convention on Climate Change (UNFCCC).

[9]World Bank, *State and Trends of the Carbon Market 2010* (Washington, D.C.: May 2010).

uncertain, in part because the CDM's screening process could not fully ensure offset quality.[10]

There are also "voluntary" carbon offset programs, where purchasers do not face legal requirements to limit emissions but may buy offsets for various reasons. For example, companies may purchase offsets to demonstrate their environmental stewardship, while individuals may purchase offsets to compensate for emissions resulting from their personal travel or consumption of fossil fuels. Because the federal government has not adopted binding limits on greenhouse gas emissions, domestic purchases of carbon offsets generally fall within the voluntary portion of the market. Voluntary programs in the United States include private sector programs, such as the Climate Action Reserve (CAR) and the Voluntary Carbon Standard (VCS), as well as Climate Leaders, an industry-government partnership overseen by EPA. Voluntary offset programs represent a relatively small share of the offset market—in 2009, the total value of the voluntary offset market was approximately $338 million, around one-eighth of the CDM market.[11] Our prior work on U.S. voluntary markets suggests that many quality assurance mechanisms exist but the extent of their use is uncertain.[12] Table 1 lists the compliance and voluntary programs we reviewed.

---

[10]GAO-09-151.

[11]World Bank, *State and Trends of the Carbon Market 2010* (Washington, D.C.: May 2010). Data on voluntary market provided by Bloomberg New Energy Finance, Ecosystem Marketplace.

[12]GAO-08-1048.

Table 1: Descriptions of Offset Programs and Standards GAO Reviewed

	Description	Offsets Issued (million tons of carbon dioxide equivalent)	
		Cumulative	In 2010
Compliance programs			
Clean Development Mechanism (CDM)	Established by the Kyoto Protocol to the United Nations Framework Convention on Climate Change (UNFCCC), the CDM enables nations with binding emissions targets under the Protocol to purchase offsets from projects in developing nations without binding targets. The mechanism is overseen by the CDM Executive Board (http://cdm.unfccc.int).	536	132
Regional Greenhouse Gas Initiative (RGGI)	Created in 2005 and implemented in 2009, RGGI regulates the carbon dioxide emissions of large fossil fuel electricity generators in 10 participating northeastern and mid-Atlantic states. Under the RGGI Model Rule, electricity generators can generally use offsets to meet 3.3 percent of their compliance reduction (www.rggi.org).	0	0
Voluntary programs or standards			
Climate Action Reserve (CAR)	A voluntary offset program that establishes standards for the development, quantification, and verification of offset projects in North America (www.climateactionreserve.org).	10.5	7.9
Chicago Climate Exchange (CCX)[a]	A voluntary greenhouse gas reduction and trading system through which members made commitments to decrease their emissions. CCX participants could trade offsets generated from qualifying emissions reduction projects (www.chicagoclimatex.com).	83.5	1.5
Climate Leaders[b]	An EPA industry-government partnership where EPA has provided technical assistance to companies on how to calculate and track greenhouse gas emissions over time, calculate emissions reductions from offsets, and incorporate offsets into emission reduction strategies (www.epa.gov/climateleaders).	0.012	0.009
The Gold Standard	Certifies projects in the voluntary market, and offers an additional quality "label" for projects that have already been approved through the CDM. The Gold Standard focuses on renewable energy and energy efficiency projects with sustainable development benefits for the local community (www.cdmgoldstandard.org).	5.4	3
Voluntary Carbon Standard (VCS)	Initiated by The Climate Group, the International Emissions Trading Association, and the World Economic Forum in late 2005 to standardize and provide transparency and credibility to the voluntary offset market, among other objectives (www.v-c-s.org).	52	29

Source: GAO analysis of offset program documents and information provided by program officials.

[a] According to CCX officials, the part of the program involving emissions reduction commitments was discontinued in 2010. However, CCX has announced the operation of an Offsets Only program for 2011 and 2012.

[b] Climate Leaders did not register offset projects or issue offsets. Instead, the program approved the use of offset tons by Climate Leaders partners to meet emissions reduction goals, assuming those tons met program criteria.

While the project review process can vary by program, it often involves the following basic steps: (1) preparing application documents, (2) establishing that the project meets eligibility criteria, (3) approving the project and registering it in a database, (4) monitoring emissions reductions over time, (5) verifying the amount of emissions reductions produced over a certain time period, and (6) issuing offsets. Existing programs generally have an administrative body to oversee offset projects and ensure they meet established quality criteria. Other key participants include project developers, who identify and perform actions that reduce, avoid, or sequester emissions, and third-party verifiers, who ensure that projects adhere to relevant quality assurance mechanisms. Figure 1 illustrates the CDM's project cycle.

Figure 1: CDM Project Cycle

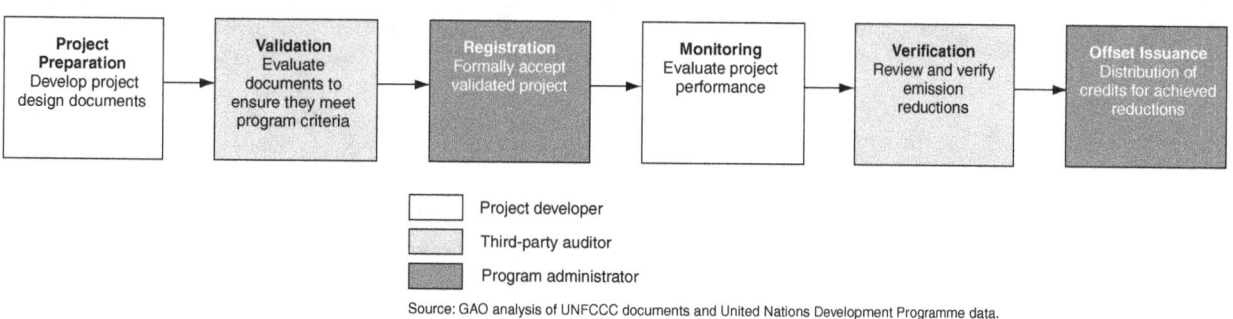

Source: GAO analysis of UNFCCC documents and United Nations Development Programme data.

Key Offset Quality Challenges

Experts and stakeholders identified five key challenges to assessing the quality of offsets in existing programs. First, many experts and stakeholders agreed that the primary challenge is assessing whether the offset project results in additional emissions reductions. Second, emissions reductions from some types of offset projects, particularly soil and forestry projects, can be difficult to measure. Third, carbon stored through soil and forestry projects may not be permanent. Fourth, in some cases it can be difficult to verify that offset projects complied with program rules and that emissions reductions occurred as expected. Fifth, the types of projects that are the most difficult to assess—forestry, international, and certain agriculture projects—may make up the majority of offsets in a future U.S. program, posing challenges for policymakers designing an offset program.

GAO-11-345 Climate Change Issues

Additionality Is the Primary Challenge

According to many of the experts and stakeholders we interviewed, the primary challenge to assessing offset quality is determining whether offsets generate "additional" emissions reductions—reductions that would not have occurred without the incentives provided by the offset program. In theory, offsets allow regulated entities to emit more while maintaining the emissions levels set established by a cap-and-trade program or other program to limit emissions. However, if the offsets represent emissions reductions that would have occurred anyway, net emissions may exceed the cap and compromise the environmental integrity of the program. We previously identified additionality as a challenge to offsets in 2008 and 2009.[13]

Although each program we examined took steps to ensure the additionality of offsets, evidence suggests that non-additional offsets have nonetheless been awarded under some existing programs. For example, the CCX, a voluntary program, awarded offsets to farmers who had practiced the credited activity for years.[14] Several studies on the CDM also suggest that a substantial number of non-additional projects have received offsets,[15] although some experts reported that the CDM has improved the quality of its offsets significantly in recent years.[16]

Experts and stakeholders cited a number of reasons why assessing additionality can be challenging, including the following:

---

[13]See GAO-08-1048, GAO-09-151, and GAO-09-456T.

[14]The farmers earned credits for conservation tillage, an agricultural practice that stores more carbon in soil than regular tillage.

[15]One study analyzed documentation from 93 projects that were registered from 2004 to 2007, and concluded that additionality was questionable in approximately 40 percent of these projects. However, the author noted that this figure was based on past performance and did not reflect recent improvements to the approval process. See Lambert Schneider, *Is the CDM Fulfilling Its Environmental and Sustainable Development Objectives? An evaluation of the CDM and options for improvement* (Öko-Institut: Berlin, 2007). Another study of 222 CDM projects concluded that approximately 26 percent of projects in the sample were likely to be non-additional. However, like the previous study, this analysis does not reflect recent program improvements. See H. W. Au Yong, *Investment Additionality in the CDM*. Technical Paper. Edinburgh, Ecometrica Press (2009).

[16]CDM officials we spoke with cited a number of recent initiatives aimed at improving offset quality while streamlining the approval process, including (1) developing further guidelines for additionality, (2) simplifying methodologies for measuring emissions by identifying superfluous requirements as well as requirements that needed further explanation, and (3) various initiatives to improve the performance and accountability of verifiers.

- *Difficulty of setting a baseline.* Assessing additionality involves comparing a project's expected reductions against a projected baseline of what would have occurred in the absence of the program. While this is not a challenge unique to offset programs—many policy decisions involve assessing alternative policies against a hypothetical baseline—it may involve a number of assumptions that are uncertain. For example, some programs approve offsets for forest management practices, such as lengthening harvest cycles to allow forests to store carbon for longer periods. An offset program could establish a baseline for these projects by assessing historical data about how forest owners respond to changes in timber prices and other economic variables. However, it may be difficult to account for the variety of decisions a forest owner may make that affect the amount of carbon stored—for example, not all forest owners may want to maximize the amount of timber produced. Assumptions regarding this and other factors that affect the amount of carbon stored can have a significant impact on the number of offsets awarded, according to some studies. For example, one study suggested that the number of offsets awarded for a hypothetical forest management project could vary by an order of magnitude, depending on the approach used to set baselines.[17]

- *Asymmetric information.* To evaluate the additionality of a project, program administrators must often rely on information provided by applicants, and in some cases, this information may be difficult to evaluate. One additionality test used by the CDM requires wind power developers, for example, to establish that a project either is not financially feasible without the revenues from offsets or is not the most economically attractive option. This can involve a complex analysis including assumptions about the internal rate of return for the project, the cost of financing, the relative costs of fuels, and the lifetime of the project. Research suggests that it can be difficult to verify these assumptions, especially since applicants know more details about the project than program administrators or verifiers, and may present data selectively to support claims of additionality.

- *Multiple incentives.* According to literature we reviewed, in some cases there may be reasons to pursue an activity that are unrelated to the offset program. For example, energy efficiency and renewable energy projects may be profitable on their own, making it difficult to gauge how offset

---

[17]Christopher S. Galik, Daniel Richter, Megan L. Mobley, Lydia P. Olander, Brian C. Murray, *Climate Change Policy Partnership: A Critical Comparison and Virtual "Field Test" of Forest Management Carbon Offset Protocols*, Duke University, October 2008.

revenue affects these projects' financial viability.[18] Similarly, conservation tillage is an agricultural practice that can earn offsets because it stores more carbon in soil than regular tillage, but farmers may also practice it for other reasons, such as to help soils retain moisture. One study suggests that conservation tillage increased by 3.5 percentage points between 1998 and 2004 as a share of total planted acres.[19] If conservation tillage offsets are accepted under a future offset program, it may be difficult to determine what portion of future increases is attributable to the offset program. In addition, some land use practices may be eligible for other federal subsidies or policy incentives outside of the offset program, potentially complicating additionality assessments.[20]

- *Misaligned incentives.* Some experts suggested that an offset program may create disincentives for policies that reduce emissions. For example, under an offset program that allows international projects, U.S. firms might pay for energy efficiency upgrades to coal-fired power plants in other nations. According to our previous work, this may create disincentives for these nations to implement their own energy efficiency standards or similar policies, since doing so would cut off the revenue stream created by the offset program. For example, some wind and hydroelectric power projects established in China were reviewed and subsequently rejected by the CDM's administrative board amid concerns that China intentionally lowered its wind power subsidies so that these projects would qualify for CDM funding. In addition, our review of the literature suggests that in some cases an offset program may unintentionally provide incentives for firms to maintain or increase emissions so that they may later generate offsets by decreasing them. This potential problem is illustrated by the CDM's experience with industrial gas projects involving the waste gas HFC-23, a byproduct of refrigerant production. Because destroying HFC-23 can be worth several times the value of the refrigerant, plants may have had an incentive to increase or

---

[18]CDM officials we interviewed said that projects that would be viable without offset revenues, such as some wind or hydroelectric power projects, could still be legitimately considered additional if a more financially attractive option—for example, a coal plant—existed. The number of credits awarded would be measured against hypothetical emissions under the most financially attractive alternative (e.g., the coal plant).

[19]National Crop Residue Management Survey, Conservation Technology Information Center. See http://www.ctic.purdue.edu/CRM/.

[20]Such incentives may include payments for protecting wetlands or preventing soil erosion issued through other government programs.

maintain production in order to earn offsets for destroying the resulting emissions.[21]

Measuring Emissions Can Be Challenging for Agricultural Soil, Forestry, and Other Types of Offset Projects

As we have previously reported, it can be difficult to accurately measure emissions from some types of offset projects, particularly soil and forestry projects.[22] An offset program needs accurate measurements of emissions to ensure that it awards an appropriate number of offsets. According to our review of the literature, the most straightforward way to measure emissions is through direct monitoring. For example, a project can run methane collected from a landfill or coal mine through a meter to measure the quantity collected and destroyed. Similarly, power plants can install monitors to measure their carbon dioxide emissions. However, direct monitoring is not feasible or cost-effective for all types of offset projects, and does not capture the effect that some projects have on emissions elsewhere. Types of offset projects with measurement challenges include the following:

- *Land-use offsets.* Land-use offset projects seek to absorb greenhouse gases or reduce emissions by affecting various natural processes. For example, trees absorb carbon dioxide from the atmosphere as they grow, and soils store carbon. However, the precise amounts stored or emitted due to an offset project may be uncertain because some of the underlying natural processes are complex and not fully understood. The amount of carbon absorbed by agricultural soils, for example, depends on the local climate, soil type, vegetation, and past land management practices. While precise methods for measuring carbon in soil samples are well established, the level of carbon will vary across a parcel of land, and changes due to the project may be small compared with the total level of carbon in the

---

[21]The CDM credits these projects based on historic baseline emissions of HCFC-22, the refrigerant of which HFC-23 is a by-product. Some research contends that refrigerant producers may have inflated their base year production levels in order to receive more offsets. (See Michael Wara, "Measuring the Clean Development Mechanism's Performance and Potential" (Stanford University, Stanford, CA: Jan 20, 2008)). A 2010 CDM Methodology Panel report was unable to state conclusively whether this had occurred, although the report recommended that the methodology be further revised to ensure that this and related issues do not occur in the future. In January 2011, member states participating in the EU ETS voted to ban CDM projects that destroy HFC-23 and nitrous oxide, although companies will be able to use credits for compliance until April 30, 2013. In a press release, the European Commission said that allowing such credits can create a perverse incentive to continue to produce or even increase production of HCFC-22.

[22]GAO-08-1048.

soil. Accurate estimates can therefore require extensive sampling, which may be prohibitively costly for some offset projects. Carbon storage projects also require ongoing monitoring to assess whether the stored carbon is re-released. According to literature we reviewed, estimates of emissions from land-use offset projects can be more uncertain than those of other projects. For example, the uncertainty of a meter that measures methane captured from a landfill may be less than plus or minus 1 percent, whereas uncertainties of the amount of carbon stored in agricultural soils range from plus or minus 6 percent to plus or minus 100 percent.[23]

- *Dispersed projects.* Offset projects that include many small sources can also be challenging to measure. For example, estimating emissions reductions from a project that distributes energy-efficient light bulbs would require assessing light bulb use among recipients and estimating the associated energy savings. According to our review of literature, one option is to collect information from a sample of recipients; however, this can cost more and may involve sampling errors or other errors compared with projects where emissions are directly monitored using a meter at a single point.

- *Projects prone to leakage.* The net effect of some types of offset projects may be challenging to measure because of the potential for emissions to increase elsewhere as a result of the project. This is known as *leakage*. For example, avoiding wood harvest in one area may simply displace harvesting and its emissions to another location. Some studies that assessed different project types in different regions suggest that leakage may be significant, although there is considerable uncertainty about the extent of leakage and the factors that cause it. Estimates suggest that between none or almost all of the emissions reductions from some types of land-use offset projects could be negated by increased emissions

---

[23]The term *uncertainty* refers to a description of the range of values that could be reasonably attributed to a quantity. An uncertainty is often presented as "plus or minus" a percentage of the estimate, meaning that the actual value could be either above or below the estimate by that amount with a certain degree of confidence.

elsewhere.[24] Other types of projects may also be at risk. For example, energy-efficiency projects may save resources that are ultimately spent on activities that increase energy use elsewhere.

Some experts suggested that measurement costs can affect the viability of certain types of projects. The measurement stringency or degree of accuracy required in a program can affect the costs of offset projects and make some types of projects unviable. Some stakeholders reported that a program will need to balance the benefits of accurate measurements with the costs. Such a balance will shift over time as new techniques and approaches are developed.

Carbon Stored in Soils and Forests May Not Be Permanent

As we have previously reported, projects that store, or "sequester," carbon carry the risk that the stored carbon will be re-released into the atmosphere, known as a reversal.[25] The risk of reversal is most commonly associated with projects involving forestry and agricultural soil sequestration. In these types of projects, reversals can occur as a result of human activity, such as logging or changes in tilling practices, or from natural events such as fires, storms, or insect infestations.

Addressing the risk of reversal is important because a reversal can negate the environmental benefit of the project. Carbon dioxide can remain in the atmosphere for a long time—up to thousands of years, according to the

---

[24]Specifically, researchers have used a variety of techniques to estimate leakage from different offset project types and activities that are similar to offsets. The estimates vary widely depending on a number of factors such as the geographic scope where leakage is considered, the location and type of project that is modeled, and other modeling choices. Results of the studies we examined ranged from less than 0 to 95 percent of targeted activities moving to other locations. See B. Sohngen and S. Brown, "Measuring Leakage from Carbon Projects in Open Economies: a Stop Timber Harvesting Project in Bolivia as a Case Study" *Canadian Journal of Forest Research* 34: 2004, p. 829–839; D. Wear and B. Murray, "Federal Timber Restrictions, Interregional Spillovers, and the Impact on U.S. Softwood Markets" *Journal of Environmental Economics and Management*, 47(2): 2004. 307–330; EPA, *Greenhouse Gas Mitigation Potential in U.S. Forestry and Agriculture*, EPA 430-R-05-006 (Washington, D.C.: November 2005); J. Wu, "Slippage Effects of the Conservation Reserve Program" *American Journal of Agricultural Economics*, 82 (November 2000): 979–992; Jianbang Gan and Bruce A. McCarl, "Measuring Transnational Leakage of Forest Conservation," *Ecological Economics*, 64(2):February 23, 2007: 423-432; and B. Murray, B. McCarl, and H. Lee, "Estimating Leakage from Forest Carbon Sequestration Programs" *Land Economics* 80(1):2004, 109–124.

[25]See GAO-08-1048.

Intergovernmental Panel on Climate Change.[26] In the context of an offset program, this means that a project in which trees planted in one year but destroyed 30 years later would convey a minimal environmental benefit compared to a project that captured and permanently destroyed methane emitted from a landfill.

Verifying Offset Projects Presents Challenges

According to our review of literature and interviews with experts, verification is an important aspect of an offset program because participants may have limited incentives to report information accurately or to evaluate quality. Verification involves confirming that the project complied with program rules and that estimates of emissions reductions are reasonable.[27] In most programs, a third-party auditor conducts the verification, which can involve checking that emissions reduction calculations are correct and site-visits to verify information with independent measurements and observations. The verifier may also review the assumptions underlying the assessment of additionality. According to our review of literature, verification may be challenging because sellers of carbon offsets may have little incentive to report information accurately to program administrators, and buyers may have little incentive to investigate the quality of offsets. Unlike buyers of other commodities, like oil or corn, buyers of offsets may not care about the quality of the offsets they buy and may be primarily interested in lowering their compliance costs by purchasing lower-cost offsets. This is partly because under some designs, buyers may not be liable for the quality of offsets they purchase after those offsets have been issued by a program.[28]

On the basis of our review of the literature and interviews with experts, we identified several challenges to verifying offset projects, including the following:

[26]According to the Intergovernmental Panel on Climate Change, about 50 percent of emitted carbon dioxide will be removed from the atmosphere within 30 years, and a further 30 percent will be removed within a few centuries. The remaining 20 percent may stay in the atmosphere for many thousands of years.

[27]We use the term *verification* to refer to both the initial assessment of whether a project conforms to a program's requirements, sometimes called a validation, as well as the assessment of emissions reductions calculations.

[28]Offset buyers may have an interest in the quality of offsets that they purchase if they are held liable for the quality of offsets they have purchased under a given program, often termed *buyer liability*.

- *Projects in developing countries and those involving complex measurement techniques can be difficult to verify.* Some experts and stakeholders suggested that offset projects in developing countries can be difficult to verify because of varying legal frameworks, lack of available documentation, or other reasons. For example, some verifiers reported that it is sometimes difficult to verify whether project developers have legal ownership of land used in a project. These challenges can vary considerably depending on the country hosting the project. Some verifiers noted that projects involving forestry and agricultural soils—in the United States or in other nations—can be more challenging to verify, since they often involve complex measurement methods. To verify emissions reduction claims in such projects, a verifier must assess the reasonableness of the model or estimation technique used, as well as the data used in the model.

- *Incentives and conflicts of interest may complicate verification.* Many experts and some stakeholders reported that misaligned incentives and conflicts of interest may affect the quality of verifications. In most cases, third-party verifiers are selected and paid by project developers. This may give verifiers an incentive to further the goals of the developer—earning offsets at low cost—over the goal of ensuring the quality of offsets.

- *Specifying verification criteria can be difficult.* Some stakeholders suggested that the verification criteria used in some programs have been unclear or subject to interpretation. This can make verifications difficult, as verifiers must make subjective judgments as to the reasonableness of assumptions and may interpret program guidelines differently than program administrators intend. For example, according to CDM documentation, about 7 percent of projects authorized by third-party verifiers in 2009 were subsequently rejected by the board that ultimately approves CDM projects. According to one study, this is partly because the CDM rules for additionality were unclear or ambiguous, which led to different interpretations between third-party verifiers and the CDM board.[29] In addition, the CDM's guidelines do not establish a level of confidence required in a verification, known as a materiality threshold. Two verifiers we interviewed suggested that that without such a threshold, verifiers may spend considerable effort investigating potential errors that would have a negligible or no impact on emissions reduction estimates.

---

[29]Lambert Schneider and Lennart Mohr, *2010 Rating of Designated Operational Entities (DOEs) Accredited Under the Clean Development Mechanism (CDM)*, Report for World Wildlife Fund, (Berlin: Öko-Institut, July 28, 2010).

- *Competence and supply of verifiers may be inadequate.* Some stakeholders we interviewed suggested that there has been a limited supply of qualified verifiers. Following spot checks of some verifiers, the CDM suspended four verification firms from 2008 to 2010, in part because of concerns over the skills and experience of staff.[30] Two stakeholders said that the shortage of verifiers is especially acute in developing countries or for more technically demanding project types such as avoided deforestation.[31] The CDM has taken various steps to improve its verification system, and these challenges may be alleviated in the future as verifiers and program administrators gain experience with the verification process.

These challenges have raised verification costs, according to our review of literature and stakeholders we interviewed. One stakeholder said that verification can be the single largest cost of developing an offset project. According to information collected by the CDM, costs range from $13,000 to $54,000 to initially register a project and $7,900 to $32,000 to periodically verify emissions reductions in that program.[32] According to two stakeholders involved in verifying CDM projects, these issues have driven up verification costs in the CDM and contributed to a growing backlog of projects. Verification costs could cause some otherwise high-quality offset projects not to be undertaken because they are not financially viable.

The Most Plentiful Types of Projects May Also Be the Most Challenging to Assess

Experts and stakeholders generally agreed that for some types of offset projects, quality is relatively easy to assess. In particular, many suggested that projects that have one emissions source and involve the metered destruction of greenhouse gases—such as methane flaring from landfills

---

[30]Third-party verification firms must be accredited by the CDM.

[31]Avoided deforestation projects aim to preserve forestlands by establishing contracts, easements, or other legal instruments to ensure that a site is not cleared of its timber.

[32]CDM requires one verification, called a validation, when an offset project is approved and registered, and a verification of the resulting reductions before offsets are issued. For comparison, the median registered CDM project expects to receive 213,000 tons of offsets through 2012. In 2009, the average price of CDM offsets was $16.6, and at those prices, the median registered CDM project would generate about $3.5 million in offset revenues. (See World Bank, *State and Trends of the Carbon Market: 2010* and United Nations Environment Programme Risoe Centre, *CDM Pipeline Overview* (Denmark: http://cdmpipeline.org, downloaded Jan. 24, 2011). However, CDM officials noted that verification costs can be substantial for smaller offset projects.

and coal mines—generally produce high-quality offsets. These projects take place at a single location; permit easy, reliable and continuous monitoring of emissions; and are not at risk of re-releasing emissions. However, offsets from such projects were forecast to be a small portion of total offsets in recent legislative proposals.[33] Further, EPA's review of recent draft legislation suggests that the potential emissions reductions from these activities may be limited, and therefore may do little to reduce the cost of a future U.S. program to limit emissions. For example, EPA's analysis of the American Clean Energy and Security Act estimated that allowing landfill, coal mine, and natural gas system methane projects as offsets would decrease the cost of emissions by only 2 percent relative to a program without these projects.

According to our review of the literature, the types of projects that are particularly challenging to assess—including forestry, international, and some agricultural offsets—may account for the majority of offsets. In 2009, CBO estimated that most offsets under proposed U.S. legislation would result from forestry and agricultural practices, with most domestic offsets coming from the forestry sector.[34] CBO also estimated that international offsets would comprise slightly over half of all offsets from 2012 to 2050. Efforts to reduce deforestation in developing countries could be a particularly significant source of offsets, given that up to 20 percent of global greenhouse gas emissions results from tropical deforestation. However, forestry offsets pose key challenges for measurement, leakage, and permanence, and have therefore had a relatively limited role in existing offset programs thus far.[35]

---

[33]These sources may also be addressed outside of an offset program. For example, emissions from many of these sources were excluded from offsets in the American Clean Energy and Security Act, which instead regulated these sources. EPA's analysis of the American Power Act is available at:
http://www.epa.gov/climatechange/economics/pdfs/EPA_APA_Analysis_6-14-10.pdf

[34]*The Use of Agricultural Offsets to Reduce Greenhouse Gases.* Statement before the Subcommittee on Conservation, Credit, Energy, and Research, Committee on Agriculture, U.S. House of Representatives (2009), of Joseph Kile, Assistant Director for Microeconomic Studies, Congressional Budget Office. Estimates are based on the offset provisions of the American Clean Energy and Security Act of 2009 (H.R. 2454).

[35]However, policymakers internationally are now considering the inclusion of reduced emissions from deforestation and degradation (REDD) in UNFCCC climate agreements.

Several Options Could Address Key Offset Quality Challenges, but Most Involve Trade-offs

According to our review of the literature and interviews with experts, policymakers have several options to choose from in addressing challenges with offset quality, but many of these options could increase the cost of offsets and may involve other trade-offs. Nonetheless, addressing these challenges may be valuable since offsets, in principle, could substantially lower the cost of a program to limit greenhouse gases relative to the cost of a program without offsets. The extent of these savings will depend partly on the quality assurance mechanisms used to address offset quality. On the basis of our review of relevant literature and interviews with experts, we identified several options that address challenges associated with additionality, measurement, permanence, or verification. We also identified steps that could address multiple offset quality challenges at the same time. Finally, we identified four overarching principles that experts generally agreed could enhance offset quality.

Several Options Could Specifically Address Additionality, Measurement, and Other Key Challenges

On the basis of our review of relevant literature and interviews with experts and stakeholders, we identified several options to address specific challenges to offset quality. Many of these options involve trade-offs—most notably, more stringent quality assurance can increase the cost of offsets. These options are not mutually exclusive, and some experts suggested that a program will likely need to employ a combination of options depending on the type of offsets allowed under the program.

Options for Addressing Additionality

There are several options to assess additionality, although many experts we interviewed stated that it may be practically impossible to ensure that all offsets are additional at the project level. Still, all of the programs we examined included additionality as a criterion for offset approval, and all took certain straightforward steps to increase the likelihood that issued offsets are additional. For example, all of the programs we reviewed seek to accept only those projects that achieve emissions reductions beyond what is already required by law or regulation, and all require that projects be initiated after a certain date (e.g., the start date of the program). The assumption behind both of these requirements is that projects that cannot meet them were likely motivated by something other than the incentives of the offset program.

All the programs we examined also take one of two approaches to more thoroughly assess the additionality of offsets—a standardized approach or a project-by-project approach. With a standardized approach, a program establishes a standard way of assessing additionality for each type of offset project and uses it for all projects of that type. One way to do this is for a program to review comparable projects and establish a performance level or set of technologies that would be considered additional. For example, a performance level for international electricity projects might reflect the most efficient method of producing electricity that is in use in a given region. Projects that exceed that performance level would then be considered additional. Alternatively, a program could identify technologies or practices that are generally additional. For example, after reviewing current livestock manure waste management practices in the United States, CAR decided that any project that installed a system to capture and destroy methane gas from manure treatment or storage facilities could be considered additional and defined a baseline methodology for all such projects.[36] Therefore, to demonstrate additionality under CAR, a project developer simply has to show that an approved methane collection system has been installed.

In contrast, with a project-by-project approach, additionality can be assessed differently for each project—even projects of the same type—so as to consider the unique circumstances of each project. For example, CDM program documents show that livestock methane capture projects generally have to (1) conduct either an investment analysis to show that methane capture was not attractive without revenue from the sale of offsets, or demonstrate that offsets allow the project to overcome some prohibitive barriers; (2) demonstrate that methane capture is not already common practice in that area; and (3) define an appropriate baseline from which offsets would be awarded. Table 2 compares these two approaches.

---

[36]Specifically, the Climate Action Reserve found that less than 1 percent of livestock operations used methane gas collection systems and that the main reason for this was that they were not commercially attractive without offset revenues. See Climate Action Reserve, *U.S. Livestock Project Protocol V3.0* (Los Angeles, Calif.: Sept. 29, 2010).

Table 2: Comparison of Project-by-Project and Standardized Approaches to Additionality Identified by Experts

	Project-by-project	Standardized
Description	Program examines the unique circumstances of each project to assess additionality.	Program establishes an approach to assessing the additionality of each project type, which is then used for all projects of that type.
Example	Projects that can demonstrate they have lower than acceptable financial returns without revenues from offsets are considered additional (investment analysis, CDM).	Installing a system to capture and destroy methane emissions from livestock manure treatment or storage facilities is considered additional (CAR).
Programs using this approach	CDM, Gold Standard, VCS	Climate Leaders, CAR, CCX, RGGI, VCS[a]
Advantages	Flexible and can be tailored to specific circumstances, easy to update with changing conditions.	Less subjective, provides certainty for developers, may be less costly to administer.
Disadvantages	Can be more costly to administer, uncertain for project developers, subjective, may award non-additional offsets.	Not appropriate for all types of projects, needs to be updated, may exclude some projects that could generate additional offsets, may award non-additional offsets.

Source: GAO analysis of program documentation and interviews with experts.

[a]Several programs also have mechanisms to consider projects outside of their primary standardized approaches, including Climate Leaders, CCX, and VCS.

The choice of approaches to address additionality involves three basic trade-offs, according to on our review of relevant literature and interviews with experts and stakeholders:

1. *Stringency versus cost.* Regardless of the approach that is used, a more rigorous assessment of additionality can be more costly to implement and exclude some projects that could have produced additional offsets, according to some experts. Two experts we interviewed estimated that relatively lenient offset standards could mean that nearly half of issued offsets are not additional. On the other hand, these experts estimated that stringent offset standards could greatly reduce non-additional offsets but exclude a significant number of potentially additional offsets from the program.[37]

2. *Up-front costs versus lower overall administrative costs.* Some experts and stakeholders suggested that a standardized approach may

---

[37]These experts stressed that such estimates are uncertain and depend on the design of an offset program. See Peter Erickson, Michael Lazarus, and Alexia Kelly, "The Importance of Program Design for Potential U.S. Domestic Greenhouse Gas Offset Supply" (accepted for publication in *Climate Policy*, 2011).

reduce administrative costs overall but may also involve higher up-front investments than a project-by-project approach. For example, the verification to register a project can cost a project developer between $13,000 and $54,000 and can take over 250 days in the CDM's project-specific process, while the same step involves minimal cost and approximately 4 to 12 weeks under CAR's standardized approach. However, developing a standard can involve up-front costs for collecting and evaluating information to assess business-as-usual activities, and for soliciting and considering public comments on proposed standards. Although a project-by-project approach may be more expensive to operate over time, an expert suggested that it can be established more quickly and at lower initial cost. This is because the program administrator would not need to establish specific standards for assessing additionality for each type of offset project, although general offset criteria for all projects would still be needed.

3. *Flexibility versus objectivity.* While standardized approaches are more objective to implement than project-by-project approaches, they are less flexible, according to some experts and stakeholders. Some stakeholders were concerned about subjective and inconsistent decisions that have occurred in some programs that use a project-by-project approach, and these concerns would likely be reduced under a standardized approach. However, once a standardized method is established, it may allow little flexibility in assessing whether a given offset project meets the standard. This lack of flexibility might mean that some projects with the potential to generate additional offsets will be excluded, and some non-additional projects will be included.

Recognizing these tradeoffs and that the suitability of a given approach may depend on the type of offset project, many experts recommended a hybrid approach that would use elements of both project-by-project and standardized approaches, and that would be tailored to each offset project type. For example, a standardized approach may work well for project types where sufficient data on relevant industry practices are available, while a project-by-project approach may be better suited to less common project types.

Options for Addressing Measurement

According to literature we reviewed, one option to address the potential for measurement error is to require project developers to incorporate measurement uncertainty into their emissions reductions calculations, reducing the number of offsets claimed to those that can be measured with a specified degree of certainty. For example, CAR adjusts the number of offsets that can be credited to a forestry project when measurement uncertainty exceeds a certain threshold. Projects measured with high uncertainty receive fewer offsets than comparable projects measured with less certainty. Such deductions can be a significant amount of potential offsets for some types of projects—up to 15 percent for some forestry projects.[38]

Additional options exist for addressing measurement challenges due to the risk of emissions leakage, according to the literature we reviewed. At the project level, some leakage may be addressed by expanding the area of emissions monitoring—for example, for certain project types, VCS tracks local "leakage belts" surrounding the project area. However, this option does not address any emissions that shift beyond a localized region. An alternative is to expand the scale of emissions monitoring to the national or international level—for example, monitoring emissions in the forestry sector or other sectors where leakage is likely to occur. In such a system, adjustments could be made if the emissions in a given sector were higher than expected, given estimated reductions from offsets. However, it may be difficult to isolate the effect of leakage from other factors that affect emissions. While some experts characterized leakage as a particularly difficult challenge, literature we reviewed suggests that assessing the potential for leakage may help policymakers adjust emissions measurements appropriately. For example, leakage may often be driven by the need to meet agricultural and timber demands. Assessing the circumstances of the markets, regions, and countries targeted by an emissions reduction program may help provide information on how much leakage can be expected, enabling program administrators to adjust policies as needed.

Options to Address Reversals

Addressing the risk of offset reversals—which occur when carbon stored in trees or soil is subsequently re-released into the atmosphere—is critical to achieving expected reductions under a program to limit emissions, according to literature we reviewed. Developing a policy to address reversals involves deciding how long a project must continue to store carbon, and how to compensate for lost reductions in the event that stored carbon is re-released into the atmosphere.

Under existing offset programs, carbon must be stored for a certain period of time, although these "permanence" requirements vary significantly. In the voluntary offset program CAR, for example, a forestry project must

---

[38]Specifically, forestry projects that have error rates of plus or minus 20 percent.

store carbon for 100 years after offsets are issued or pay back the offset credits. In contrast, CCX required a commitment of 15 years. Given that carbon dioxide can remain in the atmosphere for anywhere between 30 years and several centuries, a longer time commitment may help improve the likelihood that offset projects convey their intended environmental benefit. On the other hand, some stakeholders suggested that extended time commitments could reduce participation from landowners and renters, who may be unwilling to commit to 100-year time frames. A CAR official we interviewed noted, however, that CAR had received nearly 140 applications for forestry projects, each of which would be subject to the 100-year commitment.

The CDM takes a different approach by issuing temporary credits for forestry activities, which can be used for compliance purposes only for a certain amount of time. Once a credit expires, the owner must replace it.[39] New temporary credits can be used to replace the expiring credits if the project owner is able to demonstrate that the carbon remains stored. According to literature we reviewed, temporary crediting avoids the need for ongoing monitoring to ensure permanence, and three experts characterized it as the best option to address reversals. However, others expressed skepticism that temporary credits would be attractive to buyers in the context of a mandatory program to limit emissions.[40] One expert, for example, suggested that temporary credits would create ongoing compliance liabilities that offset buyers would be unwilling to carry. According to one study we reviewed, alternative forms of temporary crediting could address these issues—for example, allowing the private market, rather than the administrator of the program, to set contract length to meet the different needs of market participants.

On the basis of our review of the literature and experts we interviewed, we identified several other options which, together or independently, could help ensure that carbon is stored for the specified time or otherwise accounted for:

---

[39]According to CDM officials, this process effectively assumes that the carbon has been released after a certain period but offers a way to extend the compliance value of the offset if no reversal has occurred.

[40]While the CDM allows forestry activities, the EU ETS does not allow CDM credits from these activities to be used for compliance with its emissions caps. As a result there has been little demand to date for forestry projects in the CDM, and the market for temporary credits is small.

- *Hold seller or buyer liable.* Policymakers could assign liability to either project developers (sellers) or offset buyers. In the event of a reversal, the liable party would either have to replace the offsets or face sanctions for noncompliance. The advantage of holding the seller liable, according to experts and literature we reviewed, is that the landowner has a greater incentive to avoid reversals. Flexibility is another potential advantage to this option, according to one expert—a landowner that wanted to use the land for other purposes could simply replace the offsets. However, literature we reviewed suggests the transfer of liability may have to be established through a contract or other mechanism, since land ownership can shift over time. Under the buyer liability option, the responsibility for an offset reversal shifts along with the ownership of the offset. According to some literature we reviewed, this option may give buyers a greater incentive to pursue quality offsets, and liability may be easier to enforce. However, one stakeholder we interviewed suggested that such an approach would significantly dampen program participation because potential offset buyers would be unwilling to take on this level of risk. An unexpected forest fire, for example, could create a significant and immediate financial liability for an offset owner.[41]

- *Insurance.* In the case of buyer or seller liability, private insurance markets may help address the risk of offset reversals. For example, offset owners could insure themselves through private insurance or bonds issued by a bank, and if a reversal occurs, the insurer pays for the cost of replacing the offsets. According to one expert, one advantage of this option is that some private insurance companies may be better equipped to assess risk than the federal government. However, another expert noted that, because offsets are a relatively new commodity, there may not yet be sufficient information to identify risks. This expert therefore recommended against using this option until sufficient data exist to allow a private market system to work at reasonable cost.

- *Programwide buffer pools.* A program could establish a "buffer" pool by setting aside a portion of all offsets from new projects to cover possible future reversals. For example, the VCS requires land-use projects to undergo a risk assessment for non-permanence, which encompasses risks of natural disaster, technical failure, and political instability, among others. On the basis of this assessment, a percentage of the credits is

---

[41]According to some program officials, an alternative option is to enforce liability provisions only in the case of intentional reversals, while having the program administrator take on the role of replacing unintentional reversals through a buffer pool.

withheld and put into a buffer pool for use in the event of reversal.[42] According to literature we reviewed, a programwide buffer pool can serve as a type of insurance against unanticipated reversals. However, determining the appropriate size of the buffer pool may be difficult, according to some experts. A smaller buffer pool may not provide enough protection against reversals, whereas a large buffer pool may require applicants to withhold a larger share of their offsets, potentially dampening participation in the program.

Options to Address Verification

There are three basic ways to verify offset projects. First, offset projects can be verified by independent third-party organizations. Nearly all of the programs we examined use this approach. Verifiers are generally chosen and paid by project developers, presenting a potential conflict of interest. Because of this, the programs we reviewed have various requirements governing the relationship between the verifier and the developer. For example, all require conflict of interest reviews, and some have additional requirements governing the relationship between the verifier and the developer. In RGGI, for example, verifiers may not have any other direct or indirect financial relationship with project developers. Under some programs, such as the CDM, third-party verifiers may also be liable for failing to adequately verify that emissions reductions have occurred as a result of the offset project.[43] According to many stakeholders, these and other requirements generally prevent potential conflicts of interest from affecting the quality of third-party verifications, although two experts suggested that such policies may not be sufficient.

Second, some experts suggested that a program could itself verify offset projects, either directly or by contracting with third parties. This could eliminate many potential conflicts of interest by eliminating the relationship between project developer and verifier, although this is not

---

[42]Ten percent of a project's buffer is released every 5 years if the project is reverified and has the same or lower risk profile. A periodic "truing-up" ensures that total portfolio losses over time are covered by the buffer pool.

[43]For example, under the CDM, if excess offset credits are issued based on a deficient third-party verification, and certain other conditions are met, the third-party verifier must acquire and transfer an amount of reduced tonnes of carbon dioxide equivalent equal to the excess credits issued to a cancellation account maintained in the CDM registry by the Executive Board.

done in any of the programs we examined.[44] Some stakeholders suggested that having the program select verifiers could be problematic because it could add a layer of bureaucracy and could reduce market competition, among other reasons.

Third, one expert and one project developer suggested that project developers could certify their own information if a program had strong compliance and enforcement provisions to encourage developers to report truthfully. For example, the government could conduct random spot checks or audit a sample of projects. This would eliminate verifications, but could increase the risk of fraud, abuse, and mistakes.

In addition to choosing who will verify offset projects, programs face additional challenges related to verification. Experts and stakeholders identified the following options to address these:

- *Oversight can help align incentives and improve verification.* Some experts and stakeholders stressed the need for rigorous oversight to ensure verifications are effective and meet specified goals. This could take the form of accreditation processes to select third-party verifiers and ongoing monitoring of verifications including spot checks.

- *Clearly defined guidelines and expectations can facilitate verifications.* Some experts and many stakeholders indicated that clear guidelines and expectations are important for effective verification. More specific guidance and more objective criteria can reduce the chance that verifiers and program administrators will interpret information differently.

- *Standards and training can help improve the competence and supply of verifiers.* A program can help ensure that verifiers are competent by establishing standards or a minimum set of qualifications. For example, the CDM specifies that verifiers must have a certain level of verification experience before they can serve as team leaders. Some stakeholders also reported that training can be useful, although one suggested that the private sector can develop necessary training if standards are clear enough.

---

[44]One aspect of the VCS does involve program administrators choosing verifiers. When a new methodology describing an approach to monitoring, determining a project's baseline, and other provisions is submitted for approval, it gets verified twice—once with a verifier chosen by the project developer, and a second time with a verifier chosen by program staff. This is a distinct step from verifying an individual offset project.

Other Options Could Address Multiple Challenges with Offset Quality

On the basis of our review of the literature and experts we interviewed, we identified several other options that—used in combination or separately—may help address multiple challenges to offset quality at the same time. Many of these options involve addressing the quality of the program on aggregate, rather than attempting to ensure the quality of each offset at the project level. This may be necessary because, according to a CBO study, complete quality assurance of every project would be prohibitively costly, particularly for forestry and other challenging types of offsets.[45]

Limiting the Quantity of Offsets Allowed

According to our review of the literature, one way to mitigate the negative impacts of non-additional offsets, leakage, and other quality problems is to simply limit the use of offsets in a cap-and-trade program or other program to limit emissions. With this option, the emissions reduction program would ensure that only a fixed percentage of the emissions permits could be affected by any problems with offset quality.

All existing emissions reduction programs we reviewed use this option. In the EU ETS, regulated entities are able to use CDM credits for 12 percent of their emissions cap, on average, through 2012. In contrast, a draft Senate bill would have allowed a greater number of offsets into the program—approximately 42 percent of the emissions cap during the first year of the program.[46] These percentages are based on the total emissions *cap*, not the required emissions *reduction*. As a result, such limits could mean that regulated entities could use offsets for all of their required emissions reductions, assuming a sufficient supply of offsets was available. RGGI's approach, on the other hand, limits offsets to no more than 50 percent of required reductions under the cap, which may avoid a scenario where emissions reductions were wholly dependent on offsets.[47]

Restricting the number of offsets allowed would likely increase the cost of meeting the emissions cap in an emissions reduction program. On the other hand, one expert suggested that while offsets may lower the cost of

---

[45]CBO, The Use of Offsets to Reduce Greenhouse Gases (Washington, D.C.: Aug 3, 2009).

[46]American Power Act (draft bill), §§ 721(e)(1), 722(d)(1) (available at http://kerry.senate.gov/imo/media/doc/APAbill3.pdf). In subsequent years, the offset limit would have stayed flat while the overall emissions cap would have generally declined, meaning that offsets would comprise a larger share of the cap over time.

[47]Under RGGI, each source may cover up to 3.3 percent of its total reported emissions in a compliance period with offsets. According to a state official, this 3.3 percent metric is generally equivalent to 50 percent of projected avoided emissions required by the program through 2018.

compliance, such savings are irrelevant if offsets do not represent actual emissions reductions.

Limiting the Types of Offsets Allowed

Policymakers could also choose to limit the types of projects eligible for offsets, excluding the types most likely to pose quality problems. While existing offset programs we reviewed allow a wide variety of project types, they all also impose some limits on the type of projects they accept (see table 3). In some cases, programs impose limits because of concerns over the likely quality of offsets from certain types of projects. For example, soil sequestration projects, including conservation tillage, are not permitted in the CDM because of difficulties in accurately measuring the amount of carbon that is ultimately absorbed into the soil.[48]

Table 3: Eligibility of Select Offset Project Types in Select Programs

	VCS*	CCX*	CDM*	Gold Stand.	CAR	RGGI	Climate Leaders*
Energy efficiency	•	•	•	•		•[a]	•[b]
Renewable energy	•	•	•	•			
Forestry							
Reforestation/ afforestation	•	•	•		•	•	•
Forest management	•	•			•		•
Avoided deforestation	•				•		
Industrial gases	•	•[c]	•		•[d]	•[e]	
Agricultural methane	•	•	•		•	•	•
Soil sequestration	•	•					
Coal mine methane	•	•	•		•		
Landfill methane	•	•	•		•	•	•

Source: GAO analysis of program documents and information provided by program officials.

• Project type is eligible.

*Indicates that program also accepts proposals for projects from nonapproved project types.

[a]Eligible nonelectric energy efficiency measures in the building sector.

[b]Commercial boilers, industrial boilers, and bus fleets.

[c]Only ozone-depleting substances.

[d]Only nitrous oxide from nitric acid and ozone-depleting substances.

[e]Projects that reduce emissions of sulfur hexafluoride in the transmission and distribution sector.

---

[48]According to UNFCCC officials, soil sequestration can be taken into account in forestry projects, but agricultural soil projects are not allowed.

Many experts and stakeholders suggested that project types should only be eligible if they meet key quality criteria. Experts and stakeholders generally agreed on the characteristics of projects that presented relatively few quality assurance challenges:

- Projects that represent a single, localized source of emissions are less likely to necessitate resource-intensive sampling and complicated measurement models than projects that cover large areas of land or those with multiple emissions sources.

- Projects with emissions that can be measured directly through a meter allow for relatively easy monitoring and verification and are generally not subject to leakage or reversals.

- Projects that do not receive subsidies or generate revenue on their own may be less challenging to assess for additionality, since the offset is often the only financial incentive for these activities.

- Projects implemented in the United States may be easier to verify than international projects, given that verifiers may be less familiar with the legal, political, and institutional infrastructures of other nations.

Rather than limiting an offset program to only these types of projects, however, some experts cited reasons that the government should allow some flexibility around offset types. First, the supply of offsets from easy-to-monitor, low-risk projects—such as projects to capture fugitive gases from landfills or coal mines—may be limited. Second, some types of offsets that present quality assurance challenges—such as those in the forestry sector—also present large opportunities for emissions reductions. Third, imposing higher limits on international projects relative to domestic projects could exclude many legitimate reduction opportunities, according to some experts.

Many experts and stakeholders recommended developing a list of acceptable project types carefully over time. Some of them cautioned against codifying a list of acceptable project types in legislation, instead suggesting that the implementing agency choose acceptable project types using guidance from scientific and financial experts. One expert recommended that the agency initially focus on a set of project types that are most likely to produce quality offsets using the experience of existing programs and standards, and gradually build on that list as more information is collected.

| Discounting | According to our previous work, one way to compensate for offset quality problems is to discount the value of offset credits. This could be done in one of several ways, each of which has advantages and disadvantages, according to literature and experts we interviewed: |

- *Discount all offset projects.* Challenges in quantifying offsets range from assessing additionality and setting emissions baselines to measuring and verifying emissions reductions. While ideally an offset program would have measures to address these issues, our previous work suggests that even a rigorous approval process can still allow a substantial number of offsets that do not meet quality criteria. An offset program could seek to compensate for this by estimating the percentage of offsets that do not meet quality standards in the program overall and then discounting all offsets by that percentage. For example, five offset credits could be set as equal to four emissions permits in a cap-and-trade program. The burden of the discount would be borne by offset buyers, who would then need to purchase more offset credits, or by offset suppliers, who would have to perform more emissions-reducing activities. On one hand, some experts characterized this as a relatively simple approach that may help limit the adverse effects of non-additionality or other offset quality issues. However, others suggested that determining the appropriate discount would be difficult and somewhat arbitrary, and others expressed concern that discounting would reduce the chance that additional projects would be viable.[49]

- *Discount certain project types.* This option could be used to prioritize certain types of projects over others, such as projects whose reductions are relatively easy to measure or verify. These projects would receive smaller discounts—or no discount—relative to higher-priority projects. For example, some proposals suggest applying a greater discount to forestry or international projects. However, some experts cautioned that such an approach can impede economic efficiency by reducing the overall supply of offsets or by making certain types of offsets more expensive.

- *Apply a discount before credits are issued.* Under this option, used by several existing programs, discounts are incorporated into a project's measurement methodologies before credits are issued, as a way to target projects for which measurement error, leakage, or additionality is a high

---

[49]In other words, because the incremental cost of a non-additional offset is zero (compared with the baseline), suppliers would presumably be willing to sell these offsets at relatively low prices, potentially reducing the number of additional offsets.

risk. In general, experts and stakeholders supported this form of discounting when it is possible, but some noted that leakage and additionality can be especially hard to quantify and may be better addressed through other quality assurance options.

Four Broad Principles Could Improve Quality in Any Offset Program

On the basis of interviews with experts and our review of literature, we identified four broad principles that could help guide offset program design under any approach to quality assurance:

- *Identify key goals and priorities for the program.* Identifying key goals and priorities can help guide the numerous decisions that will need to be made in designing and administering the program. In many cases, policy mechanisms designed to increase the quality of offsets may also increase their cost. As a result, some experts suggested that policymakers should define an acceptable level of uncertainty—or an acceptable level of cost— on which to base the choice of quality assurance measures. Establishing these parameters may help policymakers determine whether specific types of projects can be reliably verified within the acceptable ranges of uncertainty, taking into account existing methods and technologies.

- *Align incentives with goals.* The design of the offset program creates incentives that may or may not serve program goals. Assessing the incentives created by various program designs can inform design decisions and may help improve outcomes. For example, evaluating whether the incentives offered by the offset program overlap with other incentive programs could help policymakers determine if program adjustments— such as offset discounts or limits on project types—are needed.

- *Promote transparency.* A program might cover projects from a wide range of economic sectors and countries. Clear and transparent processes and publicly available information can enable concerned third parties to be involved in project oversight, potentially improving the quality of offsets. In addition, maintaining transparency in the development of procedures and standards can help build trust in the program and reduce uncertainty for investors.

- *Incorporate evaluation and continuous improvement into the program.* Carbon markets are relatively new and less mature than other commodity markets, and program administrators will therefore need to be able to respond to an evolving marketplace. This may include adapting to unforeseen consequences of program policies as well as incorporating new technologies and innovations that emerge over time. Experts and literature thus recommended that a program develop a process for ongoing evaluation and assessment of program policies and outcomes. For example, a program could establish an ongoing process to update the methods used to establish baselines so that they accurately reflect current conditions and technologies. According to one expert, a program could also evaluate the effectiveness of its additionality procedures by assessing whether projects that had been screened out by program policies were ultimately implemented.

As agreed with your office, unless you publicly announce the contents of this report earlier, we plan no further distribution until 30 days from the report date. At that time, we will send copies to the appropriate congressional committees and other interested parties. In addition, the report will be available at no charge on GAO's Web site at http://www.gao.gov.

If you or your staff have any questions about this report, please contact me at (202) 512-3841 or trimbled@gao.gov. Contact points for our Offices of Congressional Relations and Public Affairs may be found on the last page of this report. Individuals making key contributions to this report are listed in appendix III.

Sincerely yours,

David C. Trimble
Acting Director, Natural Resources
    and Environment

Appendix I: Scope and Methodology

This report examines (1) the key challenges in assessing the quality of different types of offset projects, and (2) options for addressing key challenges associated with offset quality if the United States adopted a program to limit greenhouse gas emissions. To address these objectives, we reviewed existing information, assessed approaches in seven offset programs, and conducted semistructured interviews with knowledgeable persons in two broad groups: *experts* (researchers, economists, and academic experts involved with designing or assessing offset programs) and *stakeholders* (individuals that directly participate in or administer offset programs).

Specifically, we assessed approaches that seven offset programs use to address offset quality. We selected programs based on their representation in relevant literature and assessed two compliance programs—the Clean Development Mechanism (CDM) and the Regional Greenhouse Gas Initiative (RGGI)—and five voluntary programs—Climate Action Reserve (CAR), Chicago Climate Exchange (CCX), Climate Leaders, Gold Standard, and Voluntary Carbon Standard (VCS). We identified and interviewed 19 stakeholders from these programs to better understand quality issues from multiple perspectives. Stakeholders we interviewed included (1) program officials, (2) verifiers, and (3) offset project developers. To select a sample of verifiers, we identified seven verification firms that worked with at least three of the seven offset programs and interviewed representatives from each. To select a sample of project developers, we selected the three U.S.-based and three internationally based offset developers that had the most projects registered with the three largest offset programs in each market. Appendix II lists the stakeholders we interviewed and their affiliations.

We also selected a nonprobability sample of 13 experts—a group that included economists, academic researchers, and specialists in ecology and law—based on their knowledge and experience in relevant areas, recommendations from knowledgeable persons including agency officials and other interviewees, and the relevance and extent of their publications. To ensure coverage and range of perspectives, we selected experts who had information about key offset types, like the agriculture and forestry sectors; came from scientific, technical, or economic backgrounds, and provided perspectives from both developing offset standards and assessing the quality of offsets. We verified our list of experts with other experts that have served on previous GAO panels focused on market-based mechanisms to address climate change to ensure that we had sufficient expertise. Appendix II lists the experts we interviewed, which included agency and international officials and researchers. We conducted a content analysis to assess experts' responses and grouped the top responses into overall themes. Not all of the experts provided their views on all issues, and we do not report the entire range of expert responses in this report. Findings from our nonprobability sample of experts and stakeholders cannot be generalized to those we did not speak to. The views expressed by experts do not necessarily represent the views of GAO. To characterize expert and stakeholder views, we identified specific meanings for the modifiers we use to quantify views, as follows:

- "Many" represents 6 to 10 experts, and 7 to 15 stakeholders,

- "Some" represents 3 to 5 experts, and 3 to 6 stakeholders.

To understand the scope of current and possible U.S. government work in carbon offsets quality assurance, we interviewed officials responsible for offset-related work at agencies identified as having important roles in either existing programs or current legislation. These agencies were Energy Information Administration, Environmental Protection Agency, Department of Agriculture, and United States Agency for International Development. To understand issues related to quality assurance in the Clean Development Mechanism (CDM), we met with officials of the United Nations Framework Convention on Climate Change (UNFCCC), which administers the CDM. We also met with officials of the German Federal Environmental Ministry to learn about quality issues in the context of the implementation of the CDM on the national level. GAO provided a summary of the contents of this report to UNFCCC and EPA officials prior to its issuance.

We conducted our work from April 2010 to February 2011 in accordance with all sections of GAO's Quality Assurance Framework that are relevant to our objectives. The framework requires that we plan and perform the engagement to obtain sufficient and appropriate evidence to meet our stated objectives and to discuss any limitations in our work. We believe that the information and data obtained, and the analysis conducted, provide a reasonable basis for any findings and conclusions in this product.

Appendix II: List of Experts and Stakeholders

Experts	John Antle, Oregon State University
	Michael Gillenwater, Greenhouse Gas Management Institute
	Alexia Kelly, Department of State
	Michael Lazarus, Stockholm Environment Institute
	Jennifer Macedonia, JLM Environmental Consulting
	Bruce McCarl, Texas A&M University
	Axel Michaelowa, Perspectives
	Brian Murray, Nicholas Institute, Duke University
	Karsten Neuhoff, Climate Policy Initiative
	Lydia Olander, Nicholas Institute, Duke University
	Gordon Smith, Ecofor
	Lambert Schneider, Öko-Institut
	Michael Wara, Stanford University
Stakeholders	We interviewed officials from the following organizations:
	Offset programs and standards
	American National Standards Institute (ANSI)[1]
	Clean Development Mechanism (UNFCCC Secretariat and German Federal Environment Ministry)
	Climate Action Reserve
	Climate Leaders (EPA)
	Voluntary Carbon Standard

---

[1]ANSI coordinates U.S. participation in the International Standards Organization's (ISO) international standard-setting process. This is the process where the climate-related ISO standards were developed, 14064-1, 2, and 3 and 14065. In the United States, ANSI is an accreditation body and accredits verifiers for different offset standards.

Project developers

AgCert International Limited

AgraGate Climate Credits Corporation

EcoSecurities

Environmental Credit Corporation

TerraPass

World Bank, Carbon Finance Unit

Project verifiers

Det Norske Veritas (DNV)

Environmental Services, Inc.

ERM Certification and Verification Services

First Environment, Inc.

Rainforest Alliance

Scientific Certification Systems

Appendix III: GAO Contact and Staff Acknowledgments

GAO Contact	David C. Trimble, (202) 512-3841 or trimbled@gao.gov
Staff Acknowledgments	In addition to the contact named above, Michael Hix (Assistant Director), Quindi Franco, Cindy Gilbert, Cody Goebel, Tim Guinane, Richard Johnson, Erik Kjeldgaard, Jessica Lemke, Susan Offutt, and Ben Shouse made key contributions to this report.

www.ingramcontent.com/pod-product-compliance
Lightning Source LLC
Chambersburg PA
CBHW081402170526
45166CB00010B/3176